SpringerBriefs in Applied Sciences and Technology

SpringerBriefs present concise summaries of cutting-edge research and practical applications across a wide spectrum of fields. Featuring compact volumes of 50 to 125 pages, the series covers a range of content from professional to academic.

Typical publications can be:

- A timely report of state-of-the art methods
- An introduction to or a manual for the application of mathematical or computer techniques
- A bridge between new research results, as published in journal articles
- A snapshot of a hot or emerging topic
- An in-depth case study
- A presentation of core concepts that students must understand in order to make independent contributions

SpringerBriefs are characterized by fast, global electronic dissemination, standard publishing contracts, standardized manuscript preparation and formatting guidelines, and expedited production schedules.

On the one hand, **SpringerBriefs in Applied Sciences and Technology** are devoted to the publication of fundamentals and applications within the different classical engineering disciplines as well as in interdisciplinary fields that recently emerged between these areas. On the other hand, as the boundary separating fundamental research and applied technology is more and more dissolving, this series is particularly open to trans-disciplinary topics between fundamental science and engineering.

Indexed by EI-Compendex, SCOPUS and Springerlink.

More information about this series at http://www.springer.com/series/8884

João M. P. Q. Delgado · Fernando A. N. Silva ·
António C. Azevedo · Ariosvaldo Ribeiro

Salt Damage in Ceramic Brick Masonry

 Springer

João M. P. Q. Delgado 🆔
CONSTRUCT-LFC, FEUP
Faculty of Engineering
University of Porto
Porto, Portugal

António C. Azevedo 🆔
CONSTRUCT-LFC, FEUP
Faculty of Engineering
University of Porto
Porto, Portugal

Fernando A. N. Silva 🆔
Civil Engineering Department, Instituto
Federal de Ciências de Educação e
Tecnologia de Pernambuco
Universidade Católica de Pernambuco
Boa Vista - Recife, Brazil

Ariosvaldo Ribeiro
Departamento de Engenharia Civil, Instituto
Federal do Sertão Pernambucan
Universidade Católica de Pernambuco
Petrolina, Brazil

ISSN 2191-530X   ISSN 2191-5318   (electronic)
SpringerBriefs in Applied Sciences and Technology
ISBN 978-3-030-47113-2   ISBN 978-3-030-47114-9   (eBook)
https://doi.org/10.1007/978-3-030-47114-9

This Springer imprint is published by the registered company Springer Nature Switzerland AG
The registered company address is: Gewerbestrasse 11, 6330 Cham, Switzerland

Preface

Salt damage can affect the service life of numerous building structures, both historical and contemporary, in a significant way. Salt-induced damage represents a serious problem which a significant number of buildings must face. Besides other degradation mechanisms such as temperature changes, mechanical erosive actions of wind and water or water phase changes, salt-induced damage may have both chemical and physical nature that increases the effect of damage. It is very difficult to quantify its impacts because of lack of precise data. However, it is estimated that only the American and British transportation structures such as roads or bridges require approximately \$450 billion and £616.5 billion, respectively, to repair the damages caused by salts. The situation in the building sector would probably be correlated, requiring hundreds of billion dollars a year as the repair and maintenance costs. Soluble salts can penetrate into buildings easily with moisture which can further transport them. Therefore, understanding of moisture transport processes in porous building materials is essential to prevent salt-induced damage.

This work discusses the effects of soluble mineral salts on ceramic brick masonry walls at Petrolina, Pernambuco, Brazil, a city located 780 km from the ocean. To understand this phenomenon, a mapping of the pathologies originating from the effects of soluble mineral salts in Petrolina was carried out and wells were implemented to monitor the underground water supply in five points considered to be where the most frequent occurrence of the phenomenon takes place. Samples of soil, groundwater, bricks affected by the phenomenon and the level of chloride in the atmosphere of these localities were collected and analysed in the laboratory in order to characterize their properties. The results obtained indicate that the patho-logical manifestations present in ceramic block walls of buildings located in the study areas are influenced by a high content of soluble salts observed in the soil and groundwater samples collected, with no verifiable influence of existing chloride in the atmosphere.

Porto, Portugal   João M. P. Q. Delgado
Boa Vista - Recife, Brazil   Fernando A. N. Silva
Porto, Portugal   António C. Azevedo
Petrolina, Brazil   Ariosvaldo Ribeiro

Contents

Chapter 1
Introduction

1.1 Motivation

In Brazil, the accelerated increase in population during the last decades has contributed to the increase of the index of housing deficit. Taking into account the needs
of the general population, employment opportunities have reached low levels in rural
areas and are highest in urban centres. With the greatest concentration of industries
in urban areas, a migration of the rural population to the city seeking employment
opportunities has occurred, naturally leading to the need of an increased number
of residential units, which progressively strengthened the growth of the real estate
market (Tavares 2004).

With the growth of the real estate market in large and medium urban centres, the
prime spaces available for new developments have been reduced, causing an increase
in demand for the few existing areas. This tends to increase competitiveness due to
the scarcity of supply, especially in large metropolitan areas as well as in medium-
sized cities, generating a need to take advantage of the few existing spaces available,
forcing entrepreneurs to seek out land that is often not suitable for new ventures,
which is an already-perceived natural tendency. Most of the available land is located
in areas with no adequate drainage, with foci of soluble salts present on the surface,
requiring special treatment and specific soil analysis studies, before planning new
projects.

The scenario described above applies very appropriately to the city of Petrolina,
located in the outback of the São Francisco River, in the interior of the State of
Pernambuco, where land of these characteristics is evident in places destined for
new enterprises.

The general objective of this research is to seek a rational understanding concern-
ing the effect of soluble mineral salts in ceramic brick masonry in buildings located
within the municipal area of Petrolina.

J. M. Delgado et al., *Salt Damage in Ceramic Brick Masonry*,
SpringerBriefs in Applied Sciences and Technology,
https://doi.org/10.1007/978-3-030-47114-9_1

1.2 Objectives

The pathological manifestations present in a significant number of the ceramic block masonry walls of buildings in the urban perimeter of the municipality of Petrolina reveal the presence of soluble mineral salts.

The most frequent occurrences are found in buildings located in the lower sectors of the municipal macro drainage area.

As specific objectives, this work proposes to:

- Identify the most frequent types of pathological manifestations in masonry walls of ceramic blocks in the municipality of Petrolina, due to the action of soluble mineral salts;
- Analyze the chemical composition of the water collected in the basement of the Petrolina municipality;
- Install monitoring wells for the study and characterization of the physical and chemical composition of soil and groundwater;
- Analyze the types of soluble mineral salts in bricks used in buildings located close to the points adopted for research in Petrolina;
- Identify the natural agents that are present in subsoil water, soil, bricks and atmosphere that influence the appearance of pathological problems in buildings.

Chapter 2
State-of the-Art

2.1 Introduction

The materials used in the construction industry exhibit porous characteristics that allow the moisture fixation through mechanisms like capillarity, higroscopicity and condensation. The experiments reported in this paper are directly related to rising damp, one of the moisture propagation phenomena that cause further deterioration in buildings. Rising damp may be explained by the capillary migration of water from the soil through the porous network of the materials that compose building elements. This kind of damp assumes a greater expression in old buildings and even in new ones, mostly constructed in masonry, in which porous materials such as ceramic bricks, mortars and stones are used.

The presence of dissolved salt in the water that upraises through the porous network of building elements constitutes an aggravating factor for its degradation. If, on one hand, salts follow the water during its rise in liquid phase, the same doesn't occur when the water evaporates. Therefore, the salts remain in the porous structure of the building materials, and eventually crystallize after the solution has reached its supersaturation state (Guimarães et al. 2013). Salt can crystallize on the surface of the materials forming efflorescence with aesthetical consequences for the building, or inside the porous structure, inflicting great pressures over the pore walls that can disintegrate them when surpasses its mechanical resistance. Hygrothermal oscillation of the surrounding environment may promote cycles of crystallization/dissolution and potentially developing wrecking tensions in different pores in each cycle (Guimarães et al. 2012).

J. M. Delgado et al., *Salt Damage in Ceramic Brick Masonry*, SpringerBriefs in Applied Sciences and Technology, https://doi.org/10.1007/978-3-030-47114-9_2

2.2   Salts in Building Constructions

2.2.1   Salt Sources

There are several possibilities how salts can be brought into constructions: from air pollution, sea spray, chemical reactions and decompositions, reconstructions and renovations, de-icing agents or rising damp (Charola 2000; Warscheid et al. 2000; Lindqvist 2009; Doehne and Price 2010).

Sulfates, sulfites, nitrates and nitrites typically come from urban atmospheric deposition, but also from cement repairs. Additionally, nitrates can be often found in agriculture buildings, they are also contained in bird excrements which can pollute the urban buildings (Hall and Hoff 2007; Zappia et al. 1998). Chlorides represent a typical sea-site tracer which can penetrate into buildings due to rising damp, salt spray or flooding. Another reason can be found in utilization of sea water during mortars preparation or by application of de-icing salts for roads maintenance (Lindqvist 2009).

2.2.2   Physical, Chemical Effects and Salt Crystallization

Salts deposition over the surface of the building materials pores decrease the pore sizes and therefore enhance suction, affect the hygroscopic properties of materials and making them absorbing more moisture at the same external conditions. Both of these phenomena lead to increase of the moisture content in building materials.

Salts are able to change the drying behaviour of building materials. It has been found that at low relative humidity the drying rate of a brick saturated with a salt solution is much lower than the drying rate of a brick saturated with water. Also, a few paradoxes related to evaporation rates at different relative humidity were identified and explained (Voronina et al. 2014; Gupta et al. 2014).

Several salts in building materials can be formed as a result of chemical actions of acids. For instance, carbon dioxide, a major source of acidity in natural waters, is the most responsible for rock weathering (Sawdy et al. 2008). Another typical example is sulfur dioxide that dissolved in water it partly forms sulfurous acid and sulfur trioxide that forms acid as well. Both acids decompose lime and lime-mixed binders in coatings and contribute to the decomposition of calcium carbonate and forms calcium nitrate on walls (Zappia et al. 1998).

Salt damage, in literature also known as salt attack, salt crystallization or salt decay (see Doehne 2002), can exhibit itself by efflorescence, contour scaling (flaking), granular disintegration (powdering or sanding) or alveolization (honeycomb weathering). These damages are caused by salt phase transitions as the result of moisture transfer. Dissolved salt transported in water is deposited when the liquid is supersaturated which may easily happen in materials with high internal surface area (Moriconi et al. 1994). This often happens when the moisture transport mechanism turns from liquid capillary transport to water vapour transport, because the salts cannot be transported in a gas phase. During salt crystal growth, high stresses can arise even in large pores (Putnis et al. 1995). The crystallization pressure is higher in materials with small pores as they can better maintain the supersaturation. Other damage mechanism may be employed as well, such as hydration pressure, different thermal expansion, or osmotic pressure but the crystallization pressure is the most important anyway (Scherer 1999, 2004). It is affected by the characteristics of building material (pore structure), solution (viscosity, surface tension, vapour pressure), salt type and environment (temperature, relative humidity). A more complete list of such factors was described by Doehne (2002).

The damage caused by NaCl and other salts has a different mechanism: the salts produce an irreversible dilation during drying of the specimen and crystallization of the salt as a layer on the pore walls. No supersaturation appears to be reached by NaCl when crystallization occurs, among other things, due to solubility slightly dependent on temperature changes (Pel et al. 2002 and Lubelli et al. 2006).

2.2.3 Salt Damage Treatments

Basically, there are two main principles of desalination of building constructions (Brito et al. 2013). The first principle called passive techniques covers an environmental control, a reduction of transported moisture into a construction or a conversion of contained salts to less soluble ones and hence less damaging. The second principle actively reduces the amount of contained salts by their transportation away from the zone of deterioration or preferably completely from the whole construction. However, before application of any salt damage treatment, the pre-treatment investigation should be carried out thoroughly, revealing the nature of the salt damage:

- Evaluation of salt deterioration problems: Presumably, none of the current models for modelling of salt and moisture transport processes in porous media is fully applicable to the heterogeneous aged materials found in both historical and contemporary buildings. Simulations cannot substitute rigorous site-based observations and study.
- Evaluation of the historical and cultural value: Information about the original aspect, the significance, the aesthetic, historic or cultural role of the object within its context, all original construction and restoration phases and their historical and technical relevance should be described.
- Additional investigations: Such as a survey of the liquid moisture sources, the environmental conditions and sampling for analysis.

If the salt penetration is found, based on above described evaluation, and desalination treatments must be applied once or repeatedly, the additional information should be gathered based on the in situ sampling. Their evaluation should primarily give the information about the composition of building materials and their pore structure characteristics, moisture quantity and distribution and salts type, quantity and distribution.

2.2.4   Passive Techniques

The essence of passive techniques is based on prevention of the causes of salt damage, increased moisture in particular.

Condensed water inside the walls is one of the sources of damp. During years, the initial efforts to energy savings led to implementation of thermal insulations that were waterproofed mostly. The addition of thermally insulation layer to the interior side of an uninsulated wall decreases the temperature of the masonry during the heating season in cold climates. This can increase the risk of condensation inside the wall, mainly affecting outside corners, windows, slatted roller blind housings, ceilings, and the masonry wall base on un-heated cellars. Nowadays, vapour retarders have been devised to control moisture flow by vapour diffusion into the wall structure.

The second source of moisture is represented by rising damp that building materials absorb from the ground together with ions. Due to capillary rise and evaporation, less soluble salts will therefore reach saturation earlier, resulting in a fractionation of the crystals. However, it is very difficult to predict the precipitation sequence, i.e., less soluble salts precipitate in the lowest zone, followed by the zone of the greatest damage. The upper zone remains still wet because of rising damp instigated by the most soluble and deliquescent salts. The highest zone in the masonry is dry and unaffected by salts (see Charola 2000).

Rising damp is one of many mechanisms resulting in high moisture levels in the walls base. The management of problems due to high moisture levels requires the proper identification of the moisture source and the defect responsible, before the most cost-effective solution to the problem can be determined. Rising damp may usually be controlled by adopting one or more of the following treatment techniques (see Franzoni 2014 and Guimarães et al. 2012):

[1]   Physical or Chemical Barrier or Reduce the Absorbent Area: The main objective of this technique is to create a physical or chemical barrier at the base of the affected walls to prevent rising damp. One can place waterproof barriers; applying products by injection or by diffusion to reduce the absorbent area.

  [1.1]   Physical Barrier: A watertight material (bitumen, polymer-based mortar, corrugated sheets of stainless steel or lead sheets) is inserted into the wall's buried section to prevent water from migrating to upper levels (see Fig. 2.1).

  [1.2]   Chemical Barrier: Whereas the above-mentioned techniques of creating a damp-proof cut-off consist of a physical barrier which introduces a chemical barrier. The chemicals can be introduced into the walls by diffusion or injection through holes drilled into the walls at intervals to

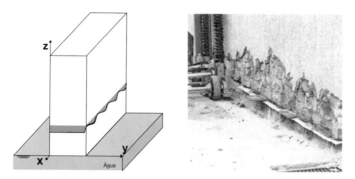

Fig. 2.1 Physical barriers—can generate structural instability

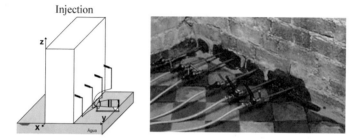

Fig. 2.2 Chemical barriers—difficulty in ensuring the continuity of impregnation across the wall thickness (commercial catalogues)

ensure that the chemical barrier covers the entire width and length of the affected walls (Fig. 2.2).

[1.3] Reducing the Absorbent Section: This technique consists basically of diminishing the absorbent area by replacing part of the wall by air, thereby not only reducing the amount of water absorbed, but also increasing evaporation (see Fig. 2.3.).

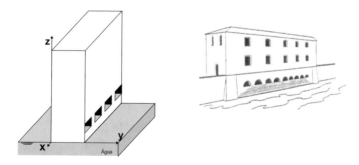

Fig. 2.3 Reduction of the absorbing section—architectural features and structural limitations

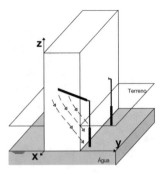

Fig. 2.4 Electro-osmotic systems—generally ineffective

[2] A Potential Against the Capillary Potential: This is an old technique that delays rising of the water by creating an electric potential against the capillary potential. This technique is no longer popular because it is not effective (see Fig. 2.4).

[3] An Atmospheric Drainage: The principle behind atmospheric drainage is the fact that damp air is heavier than dry air. Knappen believed that inserting oblique drainage tubes in walls would release damp air (coming from inside the wall), thereby facilitating the wall-drying process (see Fig. 2.5).

[4] Concealing Anomalies: When the causes of rising damp cannot be eliminated, we can decide to put up a new wall separated from the original wall with a ventilation space, which is a type of damp-proof course system concealing the anomalies (see Fig. 2.6a).

[5] A Coating with Controlled Porosity: Applying outer coatings that promote the evaporation of humidity from inside the walls and that impede salt from crystallising on the outside is a technique that conceals the problem (see Fig. 2.6b).

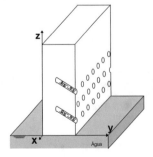 

Fig. 2.5 Aeration pipes—architectural features and efficiency limitations

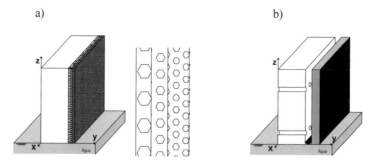

Fig. 2.6 Concealment of defects: **a** layer coating and porosity porometry controlled; **b** new sieve netting

2.2.5 Active Techniques

Active desalination techniques can be very effective for salt damage reduction, especially in cases where the damage is induced by cyclic dissolution and crystallization of salts as a response to relative humidity changes. Furthermore, no additional ingress of liquid water should be involved. According to Young and Ellsmore (2008), early signs of breaking down of a well-made and well-cured sacrificial mortars or renders may lead to the decision to proceed to desalination treatments.

There are several desalination methods available. Some of them are even at the experimental stage, therefore only the most frequently used ones are included in the presented review. These methods can be divided into the following categories:

[1] Dry-mechanical methods,
[2] Diffusive methods (baths),
[3] Methods based on diffusion and evaporation (poultices),
[4] Convective methods (by hydrostatic pressure or suction),
[5] Other methods (microwaves or biological denitrification).

There are also some methods which can be classified as outdated, therefore they should not be considered anymore. Hard cement renders, damp-proof mortar additives, Knapen tubes or passive electro-osmosis can be named as the methods belonging to this group.

The principle of dry-mechanical methods consists in removal of salt efflorescence from the surface usually by its brushing. Besides the primary removal of salts from the masonry, it can secondarily affect the thermodynamic behaviour of salts in relation to environmental conditions. Cleaner wall surface will also increase the evaporation rate which may improve the drying behaviour of the object.

Prolonged immersion and repeated or intermittent washing are the most frequently mentioned methods belonging to the diffusive category that is also called as bath methods. The principles of these methods are very similar. At first, diffusion drives the extraction of salts, following a linear dependence on time according to the Fick's law. Then, controlled by other factors such as concentration gradient, solubility or dissolution kinetics, the process slows down until the limiting value is reached. Moreover, the type of salt and the way of its distribution in the sample can also affect the whole extraction process. A repeated washing is usually more efficient except for sulfate salts and generally, the diffusive methods are the most efficient when the saltladen objects can be submerged in distilled water.

The bath methods are not used so often. The condition of the salt-laden objects to be submerged in distilled water is too restrictive. Therefore, they are applicable only for relatively small and movable objects that are not sensitive to water.

Compared to diffusion methods, salt transport and thus removal by electromigration becomes dominant over diffusion even at minimum level of applied voltage. Electromigration then continues until very low and harmless salt contents are reached, which is the main principle of this method.

On the other hand, electrokinetic method may have several disadvantages such as possible oxidation of ferrous materials and electrodes, hydrolysis and extreme pH values. The last disadvantage was eliminated by Rörig-Dalgaard (2013), who introduced a special poultice that works as a buffer component and is able to efficiently neutralize the acid from electrolysis at the anode. It is a cathode unit consisting of a three-layered poultice which neutralizes the electrochemical induced hydroxide with a buffering agent. The process is followed by precipitation of the buffering agent, citric acid, to calcium citrate inside the cathode poultice ensuring stable pH in the substrate. In addition, the precipitation of calcium citrate within the cathode poultice prevents new ions to enter the substrate during the desalination (Ottosen and Christensen 2012).

Based on previous observations, the electrokinetic method seems to work ionselectively. The best results have been achieved related to alkaline and chloride ions while the removal rate depends on the associated cation. On the other hand, only the minor part of magnesium part can be extracted. The removal rate of sulfates is also considerably slower than chlorides or nitrates, which results in gypsum precipitation inside the porous body of the materials.

Finally, there are several types of plasters according to their mechanisms of salt and moisture transport, but not all of them are suitable for desalination (see Rörig-Dalgaard 2015; Paz-García et al. 2013):

- Salt transporting systems allow the solution to migrate through the plaster and crystallize on the surface.
- Salt accumulating systems are able to absorb the salt solution from the underlying substrate and let it crystallize (accumulate) within the mortar.
- Salt blocking systems allow transport of water vapour only, while the salt solution is blocked inside the substrate.
- Moisture sealing systems ensure that both liquid water and water vapour are not allowed to cross the interface between the substrate and the mortar.

The effect of salt transportation systems is based on their pore space characteristics having the optimal structure that allows absorption of salt solution without its accumulation. It means, if the pores of the plaster are smaller than those of the salt-polluted masonry material, all soluble salts are removed from the substrate. If the substrate has a considerable amount of pores that are smaller than those of the plaster, some salt might crystallize in the plaster but a significant amount of the salt remains within the substrate itself. The accumulating systems gain their specific properties due to the presence of additives, such as water repellents. Generally, the plasters with the ability to transport the salt solution from the masonry are preferred to salt blocking systems (Petković et al. 2007).

When the plasters are able to incorporate salts in their pore space but are not resistant against their crystallization, they can be used as so-called sacrificial layers. This method is frequently used to minimize the salt deterioration of masonry although it means that used plasters have to be replaced periodically. As a typical example, lime plasters can be named, which do not reduce the evaporation and, in comparison with cement plasters, can be easily removed without remnants.

Chapter 3
Materials and Methods

3.1 Introduction

The pathological manifestations observed in a significant part of the masonry walls, of ceramic blocks, in the buildings of Petrolina rely on the presence of soluble mineral salts. The most frequent occurrences are found in buildings located in the lower areas of the macro-drainage of Petrolina.

In order to analyse this phenomenon, the mapping of the areas with the highest incidence of the action of soluble mineral salts was performed. In these locations, monitoring wells were implanted and soil and water samples were collected from the groundwater for the study. For the analysis of the chloride content in the atmosphere of the municipality in the selected localities, the Wet Candle Method was used, according to the requirements of ABNT Standard NBR 6211 (2001). After mapping the points of greatest incidence in the urban network of the city of Petrolina, based on the ABNT NBR 15495 (2007) standard, five monitoring wells were distributed in different points. Water samples were taken from the groundwater and sent to the Laboratory of Catholic University of Pernambuco (UNICAP), together with soil samples. In addition, samples of the affected and not affected bricks, by the efflorescence, were taken in the building masonry, located in the study areas. These samples were collected during four months, every 30 days.

3.2 Methodology

3.2.1 Identification of the Study Points

The identification of the locals most affected by the action of soluble mineral salts in building masonry was accomplished through visual inspection in situ. For this purpose, several surveys were carried out in each of the locations considered in the

J. M. Delgado et al., *Salt Damage in Ceramic Brick Masonry*, SpringerBriefs in Applied Sciences and Technology, https://doi.org/10.1007/978-3-030-47114-9_3

13

survey, covering a total of fifteen buildings investigated. The pathological manifestations associated with the action of soluble salts were more frequently observed in the low relief areas, in the urban perimeter of the municipality, where exist a natural tendency to open water flow and high values of moisture.

Considering the macro-drainage of Petrolina, it can be said that the city has its surface with smooth slopes towards the main receiving body. In the rainy season there is an enlargement of the streams that cut the city, due to the low rates of surface runoff (see Fig. 3.1). These facts have contributed to the occurence of floods at the lowest points of the urban network.

The situation is more serious due to the fact that most of these streams receive open sewage, mainly in the lower parts and close to the Raso da Catarina, Antônio Cassimiro, Jardim Amazonas, Dom Malan and Vila Eduardo. In this way, there is a situation of high unhealthiest in these localities.

The selection of the areas for the development of the research took into account several factors, among which the following stand out:

PM1 - Raso da Catarina

PM2 - Antônio Cassimiro

PM4 - Dom Malan

PM5 – Vila Eduardo

Fig. 3.1 Examples of open sewage in the localities of Raso da Catarina, Antônio Cassimiro, Dom Malan e Vila Eduardo

PM1—Raso da Catarina

- This is an area with a solid waste treatment station. Located between neighbour-hoods inhabited in all directions of cardinal points, with buildings in this area presenting problems arising from the action of soluble salts, appearing mainly in the neighbourhood Vila Eulalia close to Raso da Catarina. The most relevant problems are the existence of several open sewage stored inside this area, which is pumped to an aeration lagoon that is pumped again and sent through PVC pipe about 1 km poured into the open air on a canal passing downstream from the Antônio Cassimiro neighbourhood. It is worth mentioning that after this water is spilled in the open channel, it follows in the north direction about 150 m, changing direction to the west following direction to the neighbourhood Jardim Amazonas, having as a receiver the São Francisco river.

PM2—Antônio Cassimiro

- Neighbourhood located downstream of Raso da Catarina. Potentially inhabited with most of the buildings presented problems characteristic of the action of soluble mineral salts. During several surveys carried out in this area it was possible to observe the existence of sanitary sewage in the open air streets (see Fig. 3.2).

PM3—Jardim Amazonas

- Neighbourhood located in the lower area of the municipality. Totally inhabited, this neighbourhood presents buildings with walls in high decomposition stage, characteristic of the crystallization of soluble salts (see Fig. 3.3). In the area surveyed, the existence of an open-air canal was observed alongside the neigh-bourhood, whose origin is contributed by sewage from several upstream districts, including Raso da Catarina and Antônio Cassimiro.

PM4—Dom Malan

- Populated neighbourhood with an open air channel that receives input from sew-ers. Another relevant factor is the existence of a lagoon that discharges into a

Fig. 3.2 Open-air sanitary sewer in the streets being directed to the existing canal

Fig. 3.3 Building with
decaying masonry in Jardim
Amazonas

channel that moves towards the Antônio Cassimiro neighbourhood. The existing
buildings are, mostly, of medium standard and present a large number of problems
characteristic of the action of soluble salts, such as the crumbling of bricks located
near the foundation, as shown in Fig. 3.4.

PM5—Vila Eduardo

- It is considered the oldest neighbourhood of Petrolina with buildings, in the great
  majority, of average housing standard. One of the most damaging points in relation
  to the appearance of moisture in buildings is the existence of an open-air channel
  without any waterproofing, located very close to existing buildings (see Fig. 3.5).
  Even considering the existing buildings of an average standard, it is noticed that in
  the majority there are not adequate measures to inhibit the action of the possible
  soluble salts in the place, fact that could have contributed for that several buildings
  are in accelerated process of degradation in its masonry.

Fig. 3.4 Building with
decaying masonry in Dom
Malan

Fig. 3.5 Open sewage
draining near buildings

3.2.2 Characterization of the Observed Pathological Problems

The pathological manifestations present in a significant number of the ceramic block masonry walls of buildings in the urban perimeter of the municipality of Petrolina reveal the presence of soluble mineral salts. The most frequent occurrences are found in buildings located in the lower sectors of the municipal macro drainage area.

Figure 3.6 shows examples of these pathological manifestations. It is possible to observe the degradation of the ceramic bricks of different masonry walls, by the effect of soluble salts.

3.2.2.1 Methodology Applied

In order to analyse this phenomenon, mapping of the areas with the highest incidence of the soluble mineral salt effect was carried out. In these locations, monitoring wells were installed, and samples of soil and water from the water table were collected for the study. The Wet Candle Method was used to analyse the chloride content in the atmosphere of the urban area in the selected research areas.

After mapping the points of greatest incidence in the urban network of the city of Petrolina, based on the ABNT standard NBR 15495 (2001), five monitoring wells were installed at different points. At the time of installation, samples were taken from the soil located at the level of the water table. After installation, the water table was monitored, and the wells were depleted every three days for a period of nine days. Following the stabilization of the groundwater level, water samples were taken and sent for laboratory analysis together with soil samples. In addition, samples were taken of affected and non-affected bricks by efflorescence found in the masonry of buildings located in the research areas.

To determine the amount of chloride in the atmosphere, the Wet Candle Method was employed, adhering to the requirements of the ABNT standard NBR 6211 (2001).

PM2 - Antônio Cassimiro   PM3 - Jardim Amazonas

PM4 - Dom Malan   PM5 – Vila Eduardo

Fig. 3.6 Damaged ceramic bricks in masonry walls

After installation of the equipment, samples were taken every 30 consecutive days for four months.

An experimental study, in which soil, water and brick characterization tests was carried out on the affected buildings, as well as a study of the atmosphere surrounding the municipality of Petrolina-PE. Five wells for soil monitoring and five stations to capture the percentage of chloride in the atmosphere to evaluate the atmospheric salt content were installed. Each material was characterized using the demands of applicable national or international codes.

Analysis of groundwater: seventeen measurements were made, using the standardization of (APHA 2012). Soil analysis: twenty-one measurements were performed, and the methodology of the Brazilian Agricultural Research Company (EMBRAPA 1999) was used for these analyses.

Analysis of the bricks used in the masonry of affected buildings: six measurements of soluble salts were made, both on samples of damaged and non-affected bricks. For these analyses, the method adopted by the Brazilian Agricultural Research Company (EMBRAPA 1999) was used.

Determination of chloride in the atmosphere: samples were analysed for six months. The Wet Candle Method was applied according to the ABNT standard NBR 6211 (2001).

3.2.2.2 Examples of Pathological Manifestations

The identification of the locations most affected by the action of soluble mineral salts in the building masonry from Petrolina was accomplished through visual, on-site inspection. For this purpose, several surveys were carried out in each of the locations considered in the study, covering a total of fifteen buildings surveyed. The pathological manifestations associated with the action of soluble salts were more frequently observed in the low-relief areas in the urban perimeter of the municipality, where a natural tendency to open-air water runoff and perennial moisture occur.

Reviewing the macro-drainage of Petrolina, the city has a surface which gently slopes towards the main receiving body, the São Francisco River, with watercourses and streams that cut through the urban area. In the rainy season, there is an increase in water volume in the streams and brooks that pass through the city, due to slow surface run-off. These facts contribute to the emergence of flooding in the lower points of the urban network (Figs. 3.7 and 3.8).

Fig. 3.7 Degraded and powdered masonry bricks

Fig. 3.8 **a** Open-air sewage and **b** Hole executed to assess groundwater level in Antônio Cassimiro

Fig. 3.9 Brick samples location in **a** Jardim Amazonas, **b** Dom Malan neighbourhood and **c** Vila Eduardo neighbourhood

Figure 3.9 exemplifies the pathological manifestations in the research areas. The predominant characteristics are the continuous deterioration of ceramic brick masonry walls, which culminates in the fragmentation of these bricks to the point of converting them into pulverized material.

3.3   Monitoring

The monitoring wells were implemented in accordance with the recommendations of the ABNT NBR 15495 (2007). These were distributed in five sectors of the Petrolina municipal area, at points most affected by the action of soluble mineral salts. The installed wells were usual of the conventional pre-filter type. For installation, a basic project of piezometers was developed, and the parameters and procedures for implementation were defined as described below (see Fig. 3.10).

PM1—Raso da Catarina

- Monitoring well (PM1): Located in the area named Raso da Catarina. From the probe profile examination, the material found up to the depth of 2.0 m was a fine silty sand of brown colouration, and the water level was located at a depth of 1.40 meters (see Fig. 3.11).

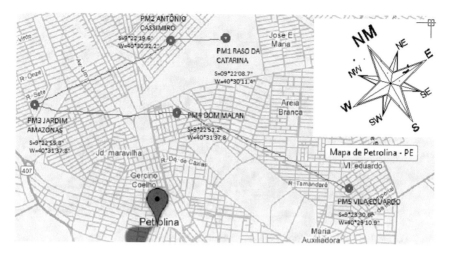

Fig. 3.10 Monitoring mapping

Fig. 3.11 Drilling of the monitoring well in Raso da Catarina

PM2—Antônio Cassimiro

- Monitoring well number 02 (PM2): Located in the Antônio Cassimiro neighbourhood downstream of PM1. According to the attached perforation profile, the soil
  characteristic is silt sand of brown colour until the depth of 40 cm, passing to
  beige colouration after this depth until reaching 1.80 m, with the water table at
  only 40 cm from the surface. The monitoring well was completed to a depth of
  1.80 and a protection lining was installed with a device used for sealing the well.
  In Fig. 3.12 it is possible to observe the monitoring well.

PM3—Jardim Amazonas

- Monitoring well number 03 (PM3): Located in Jardim Amazonas neighbourhood,
  according to the attached perforation profile, the soil is loamy sand with the

Fig. 3.12 Drilling of the monitoring well in Antônio Cassimiro

Fig. 3.13 Drilling of the monitoring well in Jardim Amazonas

existence of the water table at only 80 cm. The monitoring well was completed to a depth of 2.20 m. After reaching the water table level, a metal cap with a locking device was employed for protection (see Fig. 3.13).

PM4—Dom Malan

– Monitoring well number 04 (PM4): Located in the Dom Malan District, in this research area, there are open-air sewers and a stabilization pond near the existing buildings (see Fig. 3.14).

PM5—Vila Eduardo

– Monitoring well number 05 (PM5): Located in the neighbourhood of Vila Eduardo, as presented in Fig. 3.15, the soil is silty and silty gravel up to 40 cm, followed by silt sand up to the maximum depth reached by the well, with the

Fig. 3.14 Drilling of the monitoring well in Dom Malan

Fig. 3.15 Drilling of the monitoring well in Vila Eduardo

water table appearing at 1.00 m. The monitoring well was completed to a depth of 2.00 m with a metal cap for protection.

All the monitoring wells were executed using manual percussion drilling until the designed level was reached.

Then, an iron tube with cap and closing device was installed as a lining for protection in each of the wells, using simple concrete for fixation. The perforation records were prepared based on the data contained in the occurrence records, which were filled out at the time drilling of the wells, together with the Geo-referenced mapping of the coordinates.

Fig. 3.16 Collect and treatment of soil samples

3.4 Experimental Procedures

3.4.1 Physical-Chemical Analysis of Soil

The soil samples used in this research were extracted from all afore-mentioned study locations, and were collected at the moment of the installation of the monitoring wells. Sample collection occurred when the monitoring well reached a stable groundwater table level, using the same drilling equipment to collect the samples. After the material was collected (see Fig. 3.16), it was stored in polyethylene containers properly identified and then transported to the laboratory where they received appropriate treatment in order to avoid contamination after collection. Before being analysed, the soil was dried in disposable polyethylene trays, all labelled, and then exposed to air for three days. They were then packed in labelled, disposable polyethylene containers for later shipment to the laboratory for sample analysis.

The chemical analysis was performed using as a parameter the methodology adopted for soil analysis by EMBRAPA (1999). The physical analyses were made adopting the unified system of classification of soils using as a parameter NBR 7181 (2016). The liquidity limit was determined in accordance with NBR 6459 (2016), the plasticity limit according to NBR 7180 (2016) and the specific mass of the grains according to NBR 13602 (1996).

The samples were thoroughly analysed. Among the measurements performed, the following are the most relevant regarding the effect of soluble mineral salts in ceramic block masonry.

Fig. 3.17 Collect and treatment of subsoil water samples

3.4.1.1 Chemical Analysis of Subsoil Water

After the monitoring wells were installed in the five sites determined for testing, monitoring was begun, the water table level was checked, and the wells were depleted three days after the installation of the well (see Fig. 3.17). The purpose of these procedures, which were made every three days for ten days after implementation, was to stabilize the level of the groundwater table.

The stabilization of the groundwater table was verified, and three days later the first sample of the water was taken. The process of taking water samples from the soil used as a parameter the recommendations of the standards of ABNT, NBR 15847 (2010) and NBR 15495-1 (2007). With all the material and equipment necessary to collect the water from the subsoil, the first sample was collected and partially stored in an open polyethylene container labelled as to the volume stored, the water temperature electrical conductivity were measured. In the laboratory, the samples were analysed, having as normative support the recommendations of (APHA 2012).

3.4.1.2 Chemical Analysis of Ceramic Bricks

With the objective of analysing the ceramic bricks used in the masonry of buildings that presented signs of pathologies due to the effect of soluble mineral salts, samples were taken from damaged bricks (50 cm from the foundation) and undamaged bricks (200 cm from the foundation), at all sites where this research was carried out (see Fig. 3.18).

To analyse the bricks in the laboratory, they were ground, weighed to thirty grams and then diluted in 300 ml of distilled water into a shaken flask for twenty-four hours to obtain the sample extract. After this procedure, the anions and cations in the damaged and undamaged bricks were analysed, using the handbook of soil analysis methods of the Brazilian Agricultural Research Company—EMBRAPA (1999).

Fig. 3.18 Damaged and undamaged brick detail

3.4.1.3 Chlorides in the Atmosphere

To evaluate the chloride content in the atmosphere of Petrolina, the Wet Candle Method (see Fig. 3.19) was used according to the recommendations of NBR 6211 (2001). This test consists of a glass cylinder wrapped with surgical gauze and attached to a collection flask by a stopper. The equipment was installed on a metallic support consisting of two sheets of 50×50 cm^2 steel, one to support the damp candle and the other to cover the equipment.

After placement of the moist candle in each location, the first sample was collected after thirty days, at which time the moist candle was replaced. This process was repeated every 30 consecutive days until a total of six samples were completed after one hundred and eighty days had elapsed.

The amount of chloride in the atmosphere was analyzed according to the determinations of NBR 6211 (2001). Samples were collected at points PM1–PM5 for one

Fig. 3.19 Examples of wet candle method equipment installed

hundred and eighty consecutive days. These samples were analyzed in the Chemistry Laboratory of the Catholic University of Pernambuco. The results of the chloride content were calculated by Eq. (3.1):

$$\text{Chloride (mg/m}^2\text{d)} = \frac{2(A - B) \times M \times 25.5 V_T}{V_A \times S \times t} \tag{3.1}$$

where A is the volume of standard solution of mercury nitrate with aliquot used, in milliliters; B is the volume of standard solution of nitrate mercury with white used, in milliliters; M is the concentrate of mercury nitrate standard solution in molarity; V_T is the total sample volume after diluting, in milliliters; V_A is the volume of the aliquot taken, in milliliters; S is the exposed gauze area, in square meters and t is the exposure time in days.

Chapter 4
Effect of Soluble Mineral Salts

4.1 Introduction

In this chapter the results of the tests carried out on soil, groundwater, bricks used in buildings and chlorides in the atmosphere are presented and analyzed.

The results obtained are presented and, when applicable, a correlation between results was promoted in order to identify behavior trends. The dispersion of the measured values obtained is also presented. For a better understanding of the influence of the results in relation to the object studied, comparisons were made between the results obtained and references presented in literature.

4.2 Soil Analysis

4.2.1 Physical Properties

After collection, the soil samples were sent to the laboratory, where soil characterization was performed according to ABNT standards: Sizing Analysis, NBR 7181 (2016), Liquidity Limit, NBR 6459 (2016) and Limit of plasticity, NBR 7180 (2016).

Table 4.1 presents the results obtained, and it is possible to observe that the soils present sandy characteristics in all the studied localities: in PM-1, PM-2 and PM-5 a silt-type without plasticity, the last one 40 cm from the surface. In the other localities (PM-3 and PM-4) the soil presented clay-like characteristics. The soil of PM-5 presented, from the surface down to a depth of 40 cm, a layer of silty, loamy gravel, probably due to a nearby landfill.

Among the sites studied, the soil is easily drained subterraneously due to its sandy characteristics. Based on the results, the PM-4 and PM-3 sites present a higher tendency to retain moisture because they are characterized as loamy sand with plasticity.

J. M. Delgado et al., *Salt Damage in Ceramic Brick Masonry*,
SpringerBriefs in Applied Sciences and Technology,
https://doi.org/10.1007/978-3-030-47114-9_4

Table 4.1 Soil characterization

Local	Soil classification	Specific grain weight (g/cm)	Liquidity limit (%)	Plasticity limit	Plasticity index
Raso da Catarina	SM	2640	–	–	–
Raso da Catarina Vila Eulália	SM	2610	–	–	–
Antônio Cassimiro	SM	2670	–	–	–
Jardim Amazonas	SC	2510	28.65	19.71	8.94
Dom Malan	SC	2560	19.43	15.30	4.13
Vila Eduardo	GM SM	2700 2640	18.25/ 0.00	15.30/ 0.00	3.21/ 0.00

SM—Silty sand, SC—Clay sand and GM—Silty clayey gravel

4.2.2   Chemical Properties

Table 4.2(a) and (b) summarizes the results of the various analyses performed on the collected soil samples. Among observed findings, the following are the most relevant for the effect of mineral salts soluble on ceramic brick masonry.

The exchangeable soluble calcium cations (Ca^{2+}) and exchangeable magnesium (Mg^{2+}) appear with values ranging from 0.05 to 23.86 cmolc/kg, which is not considered elevated in relation to the other elements analyzed, a fact which later justified comparing the respective results with Motta and Ferreira (2011), where it shows that the values these elements do not increase when flooded with raw sewage, an element found in all points adopted for this study, nor with manure, a predominant element in PM-1.

The exchangeable potassium cation (K^+) appears at its highest level in PM-1 (27.6 cmolc/kg), the readings of which are lower in the other locations, a fact that is justified by the influence of leachate on PM-1, nonexistent in the other points adopted for the research, taking into consideration the data of Motta and Ferreira (2011), which shows an increase in potassium in natural soil after being flooded with slurry. In light of Table 4.3 presented by Motta and Ferreira (2011), it can also be stated that the raw sewage present at the surveyed sites does not influence the increase of potassium in the soil. However, Feitosa (2009) shows that soil mixed with 15% sewage sludge increase the potassium content considerably.

The exchangeable sodium (Na^+) in PM-3 presented a higher value (250.25 cmolc/kg), in relation to the other sites surveyed. Based on Motta and Ferreira (2011), a significant increase of this element can be seen in the soil when flooded with raw sewage, surpassing that of the natural soil which initially presented 0.02 cmolc/kg to 0.7 cmolc/kg of exchangeable sodium after flooded with raw sewage. This fact is

Table 4.2 (a) Results of the analyses of soil samples from the municipality of Petrolina-PE. (b) Results of the analyses of soil samples from the municipality of Petrolina-PE

N°	Measurement	PM1 Raso da Catarina	PM2 Antônio Cassimiro	PM3 Jardim Amazonas
01	Fine sand, tf (%)	89.62	99.58	74.99
02	pH in water (−)	7.75	7.38	7.84
03	pH in KCl (−)	7.55	6.84	6.84
04	pH in $CaCl_2$ (−)	7.63	7.01	7.63
05	$Ca^{2+} + Mg^{2+}$ exchangeable ($cmol_c$/kg)	26.66	3.03	27.93
06	Ca^{2+} exchangeable ($cmol_c$/kg)	23.86	0.05	13.30
07	Mg^{2+} exchangeable ($cmol_c$/kg^{-1})	2.80	2.98	14.63
08	Na^+ exchangeable (cmolc/kg)	39.75	36.74	250.25
09	K^+ exchangeable (cmolc/kg)	27.6	6.3	12.10
10	S value—cations sum (cmolc/kg)	94.01	46.07	290.28
11	T value—cations exchange capacity (cmolc/kg)	99.05	50.82	296.27
12	Electrical conductivity in saturation extract (mS/cm/25 °C)	4230.0	1690.0	5.670.0
13	CTC—Cations exchange capacity (cmolc/kg)	30.24	28.28	50.54
14	Cl^- in saturation extract (cmolc/kg)	68.00	26.00	86.00
15	Ca^{2+} in saturation extract (cmolc/kg)	128.30	12.00	16.00
16	Mg^{2+} in saturation extract (cmolc/kg)	12.16	2.92	9.72
17	Na^+ in saturation extract (cmolc/kg)	247.10	54.80	277.20
18	K^+ in saturation extract (cmolc/kg)	60.00	10.10	0.00
19	SO_4 in saturation extract (cmolc/kg)	276.06	12.14	2.98

(continued)

Table 4.2 (continued)

N°	Measurement	PM1 Raso da Catarina	PM2 Antônio Cassimiro	PM3 Jardim Amazonas
20	$Ca^{2+} + Mg^{2+}$ exchangeable in saturation extract (cmolc/kg)	370.00	42.00	80.00
21	Sodium saturation (%)	40.13	72.29	84.46

N°	Measurement	PM4 Dom Malan	PM5 Vila Eduardo
01	Fine sand, tf (%)	89.85	94.83
02	pH in water (−)	8.30	6.22
03	pH in KCl (−)	7.48	5.73
04	pH in $CaCl_2$ (−)	7.84	5.61
05	$Ca^{2+} + Mg^{2+}$ exchangeable ($cmol_c$/kg)	15.54	2.86
06	Ca^{2+} exchangeable ($cmol_c$/kg)	7.55	0.53
07	Mg^{2+} exchangeable ($cmol_c$/kg^{-1})	7.99	2.33
08	Na^+ exchangeable (cmolc/kg)	33.74	1.16
09	K^+ exchangeable (cmolc/kg)	17.70	4.50
10	S value—cations sum (cmolc/kg)	66.98	8.52
11	T value—cations exchange capacity (cmolc/kg)	72.31	12.83
12	Electrical conductivity in saturation extract (mS/cm/25 °C)	1610.0	281.0
13	CTC—cations exchange capacity (cmolc/kg)	35.52	36.04
14	Cl^- in saturation extract (cmolc/kg)	23.00	3.00
15	Ca^{2+} in saturation extract (cmolc/kg)	8.00	1.60
16	Mg^{2+} in saturation extract (cmolc/kg)	7.78	3.89
17	Na^+ in saturation extract (cmolc/kg)	49.80	5.70
18	K^+ in saturation extract (cmolc/kg)	2.30	4.30
19	SO_4 in saturation extract (cmolc/kg)	9.22	14.47
20	$Ca^{2+} + Mg^{2+}$ exchangeable in saturation extract (cmolc/kg)	52.00	20.00
21	Sodium saturation (%)	46.66	9.04

due to the great influence of raw sewage that this neighborhood receives from the surrounding neighborhoods situated upstream, being inferior at the other sites due to receiving lower levels of open-air raw sewage from the streets.

The electrical conductivity (EC) presented values (ranging from 281 to 4230 μS/cm at 25 °C), with higher values in PM-3 (5670 μS/cm at 25 °C) and PM-1 (4230 μS/cm at 25 °C). Motta and Ferreira (2011) show the great influence of raw sewage and leachate on the increase of EC in the soil when flooded by these two elements. It can be stated that in the case of PM-1 and PM-3, with high electrical

Table 4.3 (a) Chemical characterization of soil before and after flooding (Motta and Ferreira 2011). (b) Chemical characterization of soil before and after flooding (Motta and Ferreira 2011)

Soil type	pH H_2O	EC (µS/cm)	Na^+	K^+	Ca^{2+}	Mg^{2+}	Al^{3+}	H^+	S
			(cmolc/kg)						
Natural soil not flooded	4.19	133	0.02	0.15	0.72	2.50	0.60	2.13	3.39
Distilled water	4.42	77	0.15	0.18	0.72	1.25	0.60	1.89	2.29
Sanitary water	4.64	95	0.50	0.23	0.68	2.30	0.60	1.88	3.71
Slurry	6.83	438	1.04	0.44	0.64	1.40	2.00	1.52	3.51
Detergent	4.47	61	0.09	0.14	0.24	1.50	0.40	1.60	1.98
Raw sewage	4.50	301	0.70	0.15	0.56	1.60	0.40	1.62	3.01
Soy oil	3.64	79	0.02	0.11	0.24	1.00	2.00	10.35	1.37
Washing powder	6.83	526	2.68	0.17	0.36	1.30	2.00	1.15	4.51

Soil type	CTV (cmolc/kg)	RC (cmolc/kg)	Tr (cmolc/kg)	V (%)	m (%)	n (%)
Natural soil not flooded	6.12	44.34	68.01	55.40	15.03	0.36
Distilled water	4.78	32.14	53.14	47.94	20.74	3.05
Sanitary water	6.19	47.86	68.75	59.92	13.93	8.01
Slurry	7.03	61.26	78.15	49.95	36.28	14.74
Detergent	3.98	26.41	44.18	49.70	16.83	2.36
Raw sewage	5.03	37.89	55.89	59.84	11.73	14.00
Soy oil	13.72	37.49	152.49	10.01	59.28	0.15
Washing powder	7.66	72.32	85.09	58.87	30.53	34.96

EC—Conductivity, Na^+—Sodium, K^+—Potassium, Ca^{2+}—Calcium, Mg^{2+}—Magnesium, Al^{3+}—Aluminum, H^+—Hydrogen, S—Sum of bases, CTC—Cation exchange capacity, RC—Cation retention, Tr—Exchange capacity of clay cations, V—Degree of saturation by bases, m—Saturation of aluminium and n—Saturation of sodium conductivity (EC), these have a tendency to receive an influx of raw sewage from the neighboring circus districts through underground streams, as well as from the neighborhood itself, and may also receive some minor influx of slurry at the PM-1 locality.

The PM-2 well, according to Table 4.4, registers a value of 1690 μS/cm at 25 °C, which is lower than the ones already mentioned for PM-3 and PM-1. Compared with Motta and Ferreira (2011), presenting lower values is easily understood because this neighborhood contributes less open-air raw sewage in relation to the already mentioned sites. The main contributor is the same PM-2 well which, besides contributing to the increase of the conductivity electric power, also contributes to the downstream districts—a previously reported proven fact—when raw sewage from this neighborhood was observed being thrown in a non-waterproofed channel that flows towards the PM-3 well.

The PM-4 well has an electrical conductivity value (1610 μS/cm at 25 °C) very close to that already mentioned in the case of PM-2, the raw sewage being a probable contributor to the increase of the EC. Compared to PM-2, both having similar characteristics, the difference being that this district does not receive the leachate contribution because of being located at an adverse location in relation to PM-1, and not having any other source containing the slurry, and which offers possible conditions to send these contaminates to this site.

The values of EC (electrical conductivity) are considered high at all points, when compared to that of Prado (2005), which determines the value of 7 dS/m at 25 °C (or 7 μS/cm or 25 °C) as a limit to characterize the soil as saline. Thus, the soil in all localities is considered saline because it presents CE superior to the limit determined by Prado (2005). Since the soil is characterized as saline, it can contribute to the increase of saline deposits in the masonry, which when crystallized can affect the buildings located near the areas where the samples were taken.

The percentage of sodium saturation (%) appears with equitable values at points PM1 to PM4 (ranging from 40.13% to 72.29%). Said values are higher than those found at point PM5 (9.04%).

At points PM1 to PM4, when compared to Motta and Ferreira (2011) results, which shows the influence of the raw sewage and leachate on the increase of this element, whereas before the flood the natural soil registers 0.36%, increasing to 14.74% after flooding with raw sewage and 14% after flooding with leachate, elements which are predominate at aforementioned research sites.

It can be observed that most of the streets in these locations are not paved, a fact that facilitates the infiltration of water into the soil, contributing directly to the increase of humidity, with the exception at the point PM3 where several streets are paved. However water infiltration receives direct contributions from other upstream districts as previously mentioned.

PM5 recorded a lower value in relation to the other sites (9.04%), a fact that is due to a much lower sewage contribution than that of PM1 to PM4 considered in the research, as a result of pavement in the most of the streets at this locality, thus preventing the infiltration of water into the soil while directing it to an open channel which drains into the São Francisco river; therefore, decreasing more and more the

Table 4.4 (a) Subsoil water analysis results in Petrolina. (b) Subsoil water analysis results in Petrolina

Measurement	PM1—Raso da Catarina		PM2—Antônio Cassimiro		PM3—Jardim Amazonas	
	Samp-1	Samp-2	Samp-1	Samp-2	Samp-1	Samp-2
pH (-)	8.21	7.43	7.74	6.81	7.79	7.38
Apparent colour (mg/L PtCo)	150.0	125.0	5000.0	50.0	3500.0	80.0
Real colour (mg/L PtCo)	100.00	175.00	75.00	70.00	15.00	70.00
Elect conducti (μS/cm/25 °C)	7730.0	6390.0	8250.00	3890.0	16900.0	15300.0
Turbidity (NTU)	79.89	25.77	911.57	714.95	792.72	742.39
SD (mg/L)	5032.0	4652.0	5600.0	3.032.0	10330.0	12724.0
ST (mg/L)	5054.0	4.950.0	7782.0	4138.0	13640.0	17364.0
SS (mg/L)	22.0	298.0	2182.0	1106.0	3310.0	4640.0
SV (mg/L)	606.0	4724.0	1512.0	2746.0	2.762.0	13.354.0
SF (mg/L)	4448.0	4950.0	6270.0	4138.0	10978.0	17364.0
Na^+ (mg/L)	527.80	837.83	708.20	532.10	988.90	2.525.73
K^+ (mg/L)	460.60	620.44	43.14	30.46	4.50	17.10
Cl^- (mg/L)	814.98	1426.57	1489.96	1431.57	2964.92	6507.15
SO_4^{2+} (mg/L)	927.60	954.18	154.00	4.64	75.50	120.00
Ca^{2+} (mg/L)	300.60	320.64	280.56	200.40	621.24	761.52
Mg^{2+} (mg/L)	133.71	145.80	243.10	170.17	454.60	544.54
$CaCO_3$ (mg/L)	1300.0	1400.0	1700.0	1200.0	3420.0	4140.0

Measurement	PM4—Dom Malan		PM5—Vila Eduardo	
	Samp-1	Samp-2	Samp-1	Samp-2
pH ($-$)	7.59	7.70	6.74	5.13
Apparent colour (mg/L PtCo)	700.0	90.0	700.0	90.0
Real colour (mg/L PtCo)	25.00	85.0	15.0	12.5
Elect conductivity (μS/cm/25 °C)	6920.0	5900.0	963.0	740.0
Turbidity (NTU)	910.21	752.09	688.11	281.82
SD (mg/L)	4498.0	4954.0	404.0	590.0
ST (mg/L)	9430.0	7586.0	1772.0	1626.0
SS (mg/L)	4932.0	2632.0	1368.0	1036.0
SV (mg/L)	1618.0	5.456.0	386.0	1.224.0
SF (mg/L)	7812.0	7478.0	1386.0	1626.0

(continued)

Table 4.4 (continued)

Measurement	PM4—Dom Malan		PM5—Vila Eduardo	
	Samp-1	Samp-2	Samp-1	Samp-2
Na^+ (mg/L)	477.70	657.37	16.60	62.89
K^+ (mg/L)	27.00	24.73	14.00	7.55
Cl^- (mg/L)	1019.97	2152.37	80.00	150.17
SO_4^{2+} (mg/L)	120.75	4.14	25.10	54.08
Ca^{2+} (mg/L)	360.72	380.76	60.12	32.06
Mg^{2+} (mg/L)	230.94	279.57	48.62	24.31
$CaCO_3$ (mg/L)	1850.0	2100.0	350.0	180.0

SD—dissolved solids, ST—total solids, SS—sum of bases, SV—degree of saturation by bases, SF—fine solids

possibilities of infiltration into the soil at elevated levels. Site PM5 stands out from the other sites, as it does not receive an influx of raw sewage from other districts. Due to the location, however, an unfavorable share of runoff flows in its direction.

The percentages of sodium saturation in Table 4.2 when compared to the reference of Prado (2005), at points PM1 to PM4 the soil has a sodic classification with Na^+/CTC $\geq$ 15%. At point PM5 the soil has a solodic classification with a Na^+/CTC value between 6 and 15%.

Considering the fact that buildings need to be exempt from soluble salts and moisture, it should be noted that in the research sites the soil does not meet these characteristics and plays a significant role in the evolution of the pathological manifestations resulting from the effect of these saline elements.

The sulfate registered in PM-1, has a high value of 276.06 mg/100 g soil, a fact that, according to Metha and Monteiro (1994), characterizes the soil, as to the degree of seriousness, as a very severe assault. At the other sites, the soil reports a degree of moderate to negligible severity, because the values presented are inferior to 0,2% of sulfate composition.

Other determinations analyzed in Table 2 are not mentioned here, because they do not represent an influence on the effect of soluble mineral salts in brick masonry.

4.3 Subsoil Water Analysis

Among the ascertainments carried out, the following are the most relevant to the process of the effect of soluble mineral salts in ceramic brick masonry, accompanied by a preliminary indication of the results obtained.

Examining Table 4.4(a) and (b), it is possible to observe that saline elements are present in the groundwater in all study sites and accurately identify their influence on the effect of soluble salts in buildings located at Petrolina-PE. In view of the results obtained, the following considerations can be reported:

Table 4.5 Subsoil water analysis results in Boa Viagem—Recife, Pontes (2006)

Measurement	Boa Viagem—Recife					
	Samp-1	Samp-2	Samp-3	Samp-4	Samp-5	Samp-6
pH $(-)$	7.30	7.50	7.30	7.90	6.90	7.98
Elect conducti (μS/cm/25 °C)	45100.0	47200.0	43100.0	50000.0	52300.0	51200.0
Na^+ (mg/L)	9406.0	9509.9	10344.0	10444.0	11279.8	11586.0
K^+ (mg/L)	534.09	583.81	633.52	623.59	588.70	540.86
Cl^- (mg/L)	18434.0	18059.5	20738.2	20029.2	21092.8	20520.3
SO_4^{2+} (mg/L)	2967.0	3198.8	3532.4	3569.7	3569.7	3668.9
Ca^{2+} (mg/L)	560.0	640.0	720.0	440.0	480.0	480.9
Mg^{2+} (mg/L)	1166.4	1166.4	1166.4	1336.5	1530.9	1.61.4

- There is high electrical conductivity in the groundwater in all the points adopted for the research.
- PM-3 contains the highest value of electrical conductivity (16900 μS/cm at 25 °C), with high values being reported at PM-2 (8250 μS/cm at 25 °C), PM-1 (7530 μS/cm at 25 °C), and at PM-4 (6920 μS/cm at 25 °C), while PM-5 reports the lowest value (963 μS/cm at 25 °C).

Although the value observed at PM-5 was the lowest of all, these values still constitute a high reference since the water used in civil construction needs to be potable, containing the minimum amount of impurities. It is important to note that water can infiltrate by capillarity and, when evaporating, crystallize the salts dissolved in the water, damaging the buildings.

Based on Pontes (2006), which indicates the electrical conductivity of seawater, the highest value obtained at site PM3 (16,900 μS/cm at 25 °C) represents 39% in relation to the lowest found in seawater (43100 μS/cm at 25 °C). Among the sites PM-1, PM-2 and PM-4 the lowest expression, when compared with Table 4.5, registers 16%. While the lowest value obtained (963 μS/cm at 25 °C) at site PM-5 represents 2.2% in relation to Table 4.5.

It is possible to observe that the values obtained of the electrical conductivity when compared with seawater, indicate that the water of the samples taken is incompatible with potable water suitable for use in construction. Frequent moisture from groundwater that has high electrical conductivity has a significant influence on the effect of soluble salts in masonry and other building parts.

The alkaline metals (sodium and potassium) and the alkaline earth metals (calcium and magnesium) contained in the water may influence in the appearance of efflorescence in masonry walls of ceramic blocks.

Examination shows that, except for PM-5, all sites reported the presence of sodium, potassium, calcium and magnesium ions in relatively high amounts. Indeed, in the particular case of the sodium ions of the groundwater collected at PM-3, the water exhibited a content of this ion in sample 2 of 2525.73 mg/l which is 40 times higher than the content observed in the groundwater collected in PM-5–62.89 mg/l in sample 2. This fact confirms the values of electrical conductivity found in the groundwater at these sites.

The alkaline metals (sodium and potassium) and the alkaline earth metals (calcium and magnesium) contained in the water may influence in the appearance of efflorescence in masonry walls of ceramic blocks.

Examination shows that, except for PM-5, all sites reported the presence of sodium, potassium, calcium and magnesium ions in relatively high amounts. Indeed, in the particular case of the sodium ions of the groundwater collected at PM-3, the water exhibited a content of this ion in sample 2 of 2525.73 mg/l which is 40 times higher than the content observed in the groundwater collected in PM-5–62.89 mg/l in sample 2. This fact confirms the values of electrical conductivity found in the groundwater at these sites.

Regarding the calcium and magnesium content, the quantities observed in the groundwater at the research areas were also significant, with the highest values always observed in PM-3 761.52 mg/l and 554.60 mg/l, respectively. Regarding calcium and magnesium ions, PM-5 was the site with the lowest values of the five areas studied—60.12 mg/l and 48.62 mg/l, respectively. In the case of sulfate ions, the PM-1 neighborhood had the highest value (954.18 mg/l), which was also slightly elevated at PM-2 with a value of 154.00 mg/l—while the others showed values below 120 mg/l (see Table 4.4).

The values mentioned above, when compared with those obtained in seawater in Pontes (2006) the highest value of sodium at point PM-3 (988.90 mg/l), represents 10.4% in relation to the lowest value found in seawater. Potassium at site PM-1 (620.44 mg/l) surpasses the lowest constant value in Pontes (2006), reaching 116.2% in relation to seawater.

The highest value obtained from samples (761.52 mg/l)) at point PM-3, surpassing the lowest value found (440 mg/l) in seawater, representing 173% when compared to PONTES, 2006. At sites PM-1, PM-2, and PM-4, the values obtained are high in relation to seawater. However at point PM-5 the value obtained is relatively low in relation to the other sites.

Magnesium values obtained at sites PM-1 to PM-4 have a similar behavior to that of calcium, which is more relevant at site PM-3, representing 46.7% when compared to the lowest value of Pontes (2006). At site PM-5 the values obtained were relatively low in relation to the other sites. The results of the samples show that, in the same way that contamination of the raw sewage and leachate in the soil was already indicated, the same results are repeated in relation to the sodium and potassium contents in the water at the same locations, thus proving that the origin stems from the presence of sewage and slurry.

With regard to sulfate, calcium and magnesium, the existence of sewage and leachate at the site does not necessarily indicate they are the source of contamination that influences the elevated levels of these elements in the soil. However, when compared with seawater, if the water reports high calcium, magnesium and sulfate contents, this indicates that the existence of these elements can contribute to the formation of soluble salts as a result of these higher levels of these elements.

The analysis of the previous Table 4.4 leads to the conclusion that the main ions causing crystallization presented values far above those usually found in groundwater, especially sodium, potassium, calcium and magnesium at all collection sites, except at PM-5 where the quantities reported are close to acceptable values.

With regard to the amount of chloride ions, Table 4.2 shows the results observed in each of the locations where groundwater was collected. PM-3 had the highest concentration of this ion, 6507.15 mg/l.

It is important to note that salts such as calcium chloride and magnesium chloride are very soluble and the presence of this ion in great quantity favors the appearance of the phenomenon of efflorescence in masonry walls of buildings of the locations mentioned.

Regarding the total hardness, the groundwater of the studied locations reported values considered high (see Table 4.4) that characterize them as hard water in PM-1, PM-2, PM-3 and PM-4, and reporting lower values in PM-5.

It is important to note that, due to these results, it can be concluded that the PM-5 collected groundwater presents a low salinity when compared to the other localities, an aspect that is characterized as a positive factor since the pH of the waters of this area showed an acidic characteristic. If the amount of salts were high, this locality would present favorable conditions for a strong occurrence of efflorescence in the masonry of buildings at this site, aspect not observed in the field inspections. In addition, it can be concluded from the results of the analyses that the site in which the groundwater presented the highest salinity was PM-3.

The pH values in Table 4.1 show that of the five sites investigated, four showed groundwater with compatible characteristics in a slightly alkaline medium (pH > 7.0). The only locality that presented groundwater with acidic characteristics was PM-5 (pH < 7.0). Since the acidity of a medium increases the solubilization of alkaline salts, the groundwater collected in the PM-5 presents a factor that most certainly contributes significantly to the occurrence of efflorescence in the masonry walls of ceramic blocks in buildings in this locality.

Pontes (2006) presents the values obtained in sample of seawater whose origin was Boa Viagem Beach (see Table 4.6), which when compared to that in Table 4.4 are presented as follows:

Table 4.6 Water sea composition in Boa Viagem—Recife, Pontes (2006)

Ions	Ions quantity (mg/L)
SO_4^{2+}	2800
Mg^{2+}	1300
Ca^{2+}	400
Cl^-	19900
Na^+	11000
Ka^+	400
pH	>8

- SO_4^{2+} in PM-1 (PM1), the value of 954.18 mg/l represents 34.2% in relation to seawater data, at the other points this reaches a maximum value of only 5.5%.

The sulfate found at PM-1 showed an elevated value of 954.18 mg/l, and 154.00 mg/l at PM-2, with lower values at the other sites, a fact that according to Metha and Monteiro (1994) characterizes the underground water found in PM-1 and PM-2, in terms of degree of seriousness, as a severe aggression. In the other locations, the soil presents a degree of moderate to negligible severity, because these report values lower than 150 mg/l of sulfates in its composition.

Considered the values presented in Table 4.7 (Bauer 2007), the groundwater at PM-1 presents a degree of strong aggressiveness in concrete, exceeding 600 mg/l.

- Mg^{2+} at PM-3, the most representative value (544.54 mg/l), reaching 41.88% in relation to seawater, while at the other points reaches a maximum value of 279.57 mg/l in PM-4 (21.5%) and lower value (48.62 mg/l) PM-5, representing only 1.7% relative to seawater.

Considering Bauer (2007), the PM-3 groundwater reports a strong degree of aggressiveness to concrete due to exceeding the 600 mg/l limit.

- Ca^{2+} in PM-3 exceeds the seawater value of 761.52 mg/l representing 190%, and in comparison to the seawater reference at sites PM-1, PM-2 and PM-4 presenting variations between 300 mg/l and 381 mg/l, which are percentages compatible with seawater.
- Cl^- at PM-3 reaches a higher value (6507.15 mg/l) representing 32.7% compared to seawater, while at sites PM-1, PM-2 and PM-4 the percentage vary from 4, and PM5 reaches only 0.75%. According to Nappi (2010), which indicates a limit of 500 mg/l to characterize water as saline, it can be stated that the samples are saline at points PM-1 to PM-4 and brackish at point PM-5, because it has a value below that limit.
- Na^+ at points PM-1 to PM-4 presents percentages ranging from 5.97% to 22.96% compared to seawater and only 0.6% at point PM-5.
- The K^+ at the PM-1 site exceeds the seawater reference reaching 620.44 mg/l representing 155.11%, and at sites PM-2 to PM-5 the percentages vary between 3.5% and 10.8%.

	Degree of aggressiveness		
Table 4.7 Limits degree of aggressiveness to concrete, Bauer (2007)	Weak	Strong	Very strong
pH	5.5–6.5	4.5–5.5	Lesser than 4.5
CO_2 (mg/l)	15–30	30–60	Higher to 60
NH_4^- (mg/l)	15–30	30–60	Higher to 60
Mg^{2+} (mg/l)	100–300	300–1500	Higher to 1500
$SO4^{2-}$ (mg/l)	200–600	600–2500	Higher to 2500

The pH values are very close to 8 at sites PM-2 to PM-4, reaching above 8 at PM-1 and being lower at PM-5. Considering Pontes, 2006, all values are compatible with those found in seawater in Boa Viagem Beach, which is characterized as saline water. In light of the presented values, all the characteristics found in the groundwater samples indicate that the water table as the main contributing factor of the transfer of saline elements to the buildings which, with the water evaporation, begin to crystallize and directly impact the buildings affected by these salts.

4.4 Brick Analysis

Considering that bricks are porous building materials that facilitate water penetration, the decision was made to remove samples of damaged and undamaged bricks in buildings located near sites PM-1 to PM-5 under study.

The bricks were crushed then left immersed in distilled water for 24 h, following procedures using the methodology of EMBRAPA (1999). The bricks were analyzed and it was found that in their aggregate mass there were very high values of soluble salts in the samples of damaged bricks, and relatively low in samples of undamaged bricks. Six types of water soluble chemicals were evaluated in the analyses.

From Table 4.8, the reported results that represent a greater influence towards the understanding of the research objective are.

Table 4.8 (a) Brick sample results. (b) Brick sample results

Measurements	PM1—Raso da Catarina		PM2—Antônio Cassimiro		PM3—Jardim Amazonas	
	N.D.	D	N.D.	D	N.D.	D
SO_4^{2-} (cmolc/kg)	29.39	182.66	20.89	255.80	27.36	10.93
Cl^- (cmolc/kg)	25.00	4749.87	20.00	439.99	20.00	1969.94
Ca^{2+} (cmolc/kg)	20.04	1507.01	38.08	144.29	22.04	448.90
Mg^{2+} (cmolc/kg)	34.03	211.50	12.16	44.97	8.51	127.63
K^+ (cmolc/kg)	13.81	527.47	0.00	0.00	19.22	40.85
Na^+ (cmolc/kg)	2.39	1632.38	2.39	496.83	0.00	1159.23

Measurements	PM4 Dom Malan		PM5 Vila Eduardo	
	N.D.	D	N.D.	D
SO_4^{2-} (cmolc/kg)	40.10	151.45	92.92	136.13
Cl^- (cmolc/kg)	15.00	2019.94	115.00	639.98
Ca^{2+} (cmolc/kg)	24.05	605.21	46.09	138.28
Mg^{2+} (cmolc/kg)	17.02	122.77	25.53	38.90
K^+ (cmolc/kg)	8.41	203.05	5.70	94.92
Na^+ (cmolc/kg)	0.00	922.66	56.80	828.03

N.D—Bricks not damaged and D—Bricks damaged

Sulfate reported values considered high in four samples of the damaged bricks, being higher (255.8 cmolc/kg) at site PM-2, while the other sites recorded the following: PM-1 (182.66 mg/100 g brick), PM-4 (151.45 cmolc/kg) and PM-5 (136.13 cmolc/kg), and PM-3 presented lower values in relation to the others sites. In the undamaged bricks, values were considered low, ranging from 20.89 to 92.92 cmolc/kg, when compared to those reported on damaged bricks. The higher values of sulfate in damaged bricks may have their origin linked to the contribution of agricultural fertilizers used in large scale irrigation projects, dragged to the site of the buildings where the samples were collected through the underground streams, deposited in the soil and later carried to the walls of the buildings with which it comes in contact. The sulfate values, found in the damaged bricks, were relatively low in the undamaged bricks. Silva (2011), points out that sulfate is very soluble, a fact that contributes directly to easy access to the bricks and consequent formation of crystallized salts, thus producing damage in the building materials.

The chloride in the damaged bricks presented higher results at PM-1 (4749 cmolc/kg), PM-3 (1969.94 cmolc/kg) and PM-4 (2019.94 cmolc/kg). The lowest values found in the damaged bricks were recorded in the PM-2, recording 439.99 cmolc/kg and PM-5 registering 639.98 cmolc/kg. The undamaged bricks obtained values much lower than the damaged ones, with findings between 15 and 115 cmolc/kg. Considering Silva (2011), the chloride is very soluble, which facilitates access to the buildings. The high values obtained in the samples of the damaged bricks prove that this element contributes directly to the appearance of pathological manifestations in masonry.

Calcium in the damaged bricks reported a higher value of 1507 cmolc/kg at PM1, followed by 605.2 cmolc/kg at PM4, 448.81 cmolc/kg at PM3, with lower values appearing at PM2 (144.28 cmolc/kg) and PM5 (138.27 cmolc/kg). Among the undamaged bricks, the values were much lower than those obtained in damaged bricks ranging from 20.04 to 46.09 cmolc/kg. The divergence of the amount of calcium between damaged and undamaged bricks demonstrates that the damaged bricks confirm that calcium is a soluble element (see Table 4.5) which also contributes directly to the incidence of pathological manifestations in masonry.

The amount of magnesium in the damaged bricks appeared with greater intensity at PM-1 (211.49 cmolc/kg), PM-3 (127.62 cmolc/kg) and PM-4 (122.76 cmolc/kg). At the other points, the values were lower, registering 44.97 cmolc/kg at PM2 and 38.89 cmolc/kg at PM5. Among the non-damaged bricks all the results obtained in the analyses presented lower values than those of the damaged samples, between 8.50 and 34.03 cmolc/kg. The low values presented in the soil analysis prove that soil contributes as a source of contamination of the bricks containing magnesium, and also does not rule out a possible groundwater contribution. At site PM-5, neither water nor soil present amounts of magnesium that justifies the contamination of the bricks. The fact that the source remains unknown clearly indicates the hypothesis of a connection to the raw material used for the manufacture of the bricks, which requires further study focusing on this objective.

The potassium ions in the damaged bricks presented the highest value at the PM-1 site, recording 527.47 cmolc/kg, followed by PM-4 which recorded 203.05 cmolc/kg and PM-5 reporting 94.92 cmolc/kg. Lower values appear at sites PM-2, recording 0.00 cmolc/kg and PM3 registering 40.85 cmolc/kg. The PM-1, PM-3, PM-4 and PM-5 sites presented lower amounts than those reported in the damaged bricks, ranging from 5.70 cmolc/kg to 19.22 cmolc/kg. At site PM-2, the sample of the undamaged bricks presented a value of 0.00 cmolc/kg, being equal to the value presented in the damaged brick extracted from the same location. The high amount of potassium at site PM-1 leads to the conclusion that the sample may have been influenced by the groundwater, which also reported a high level in the sample at this site due to the influence of slurry found in the area. At the other points, the lower values obtained in the samples; demonstrate that neither the groundwater nor the soil contributed to the appearance of potassium in the damaged brick, which also presented values considered to be low.

The amount of sodium found in samples of bricks from sites PM-1 to PM-4 is comparable, similar to what happens with the groundwater at the same sites, indicating the influence of the water in the contamination of bricks containing sodium. There is an exception in relation to PM-5 which presents a noteworthy sodium value in the damaged brick samples (828.03 cmolc/kg), with no significant sodium value, neither in the groundwater nor in the soil samples. The appearance of sodium in the damaged brick is of unknown origin, perhaps due to the raw material used for the manufacture of the bricks, which requires further study focusing on this objective. Samples of the undamaged bricks all showed low values when compared to damaged bricks.

Taking into consideration the data obtained, in general, soluble salts appear in the damaged bricks with elevated intensity in all the evaluated samples, being very symptomatic when compared to the undamaged bricks evaluated.

Figure 4.1 shows a comparison between the sulfate values found in the samples of damaged bricks in relation to groundwater, soil and undamaged bricks. This figure demonstrate the presence of sulfate in the groundwater, a high reading which exceeds 900 ml/l and, according to Metha and Monteiro (1994), characterizes the soil, as to the degree of severity, as moderate due to having a reading greater than 150 mg/l. The same happens with the results obtained from the samples of the damaged bricks, a fact that shows a tendency of the existing sulfate in the groundwater and the soil to have contaminated the bricks used in the building, since the trend line between the damaged bricks, soil and groundwater is balanced.

The undamaged bricks, with regard to sulfate, present little representative values for the degradation of the samples, which is to say that the bricks that came from the supplier were not contaminated with sulfate.

Figure 4.2 shows a comparison of chloride results found in samples of damaged bricks relative to groundwater, soil and undamaged bricks. This figure shows that the chloride appears with representative values in the samples of groundwater and damaged bricks, thus the tendency for the curves to approach each other. These facts lead to the conclusion that the chlorides that reach the bricks of the buildings receive greater influences from the water of the water table.

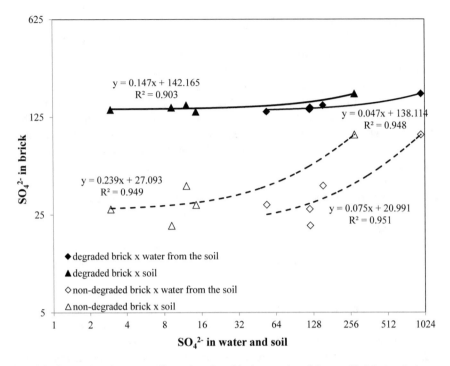

Fig. 4.1 Comparison between sulfate values found in the samples of damaged bricks in relation to groundwater, soil and undamaged bricks

The soil in this process does not report significant values when compared to those found in the groundwater. The same is true of samples of undamaged bricks, a fact which infers the bricks are exempt of chlorides when sent from factory for use in construction.

Figure 4.3 shows a comparison of calcium results found in samples of damaged bricks relative to groundwater, soil and undamaged bricks. The figure shows that calcium appears in more significant amounts in groundwater, soil and the damaged bricks, which means that the bricks are being contaminated with calcium from both the groundwater and the soil. The undamaged bricks and their values approximate the trend line in relation to the groundwater, which demonstrates the need to analyses the percentages of calcium brought from the repositories in the bricks supplied for construction.

Figure 4.4 shows a comparison of magnesium results found in samples of damaged bricks relative to groundwater, soil and undamaged bricks. The magnesium present in the soil displays values close to the trend line in relation to the damaged bricks, making it clear that the influence of the soil in the appearance of magnesium in the construction bricks, since all the areas retain humidity and present conditions favorable to the process (see Fig. 4.4). In regard to the water, although the trend line is more dispersed, some influence of the water table is not ruled out. It is clear

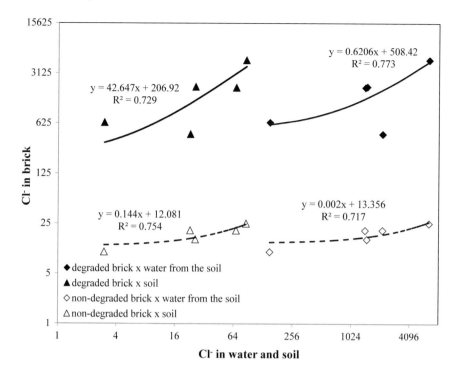

Fig. 4.2 Comparison of chloride results found in samples of damaged bricks relative to groundwater, soil and undamaged bricks

and quite easy to understand that the undamaged bricks of the trend line are not very representative, ruling out the need to analyses the deposits used by the manufacturers that supply the local market.

Figure 4.5 shows a comparison of potassium results found in samples of damaged bricks relative to groundwater, soil and undamaged bricks. The determinations of the presence of potassium in the damaged bricks are approximate in relation to the data found in groundwater and soil, thus showing a connection of the influence of both water and soil in relation to the appearance of potassium in the building bricks. It is clear that it is easy to understand that the undamaged bricks of the trend line are not very representative, ruling out the need to analyze the deposits used by the manufacturers that supply the local market.

Figure 4.6 shows a comparison of sodium results found in samples of damaged bricks relative to groundwater, soil and undamaged bricks.

The trend curve shows a closer correlation between the results obtained from the analysis of the damaged bricks and the soil, which proves that the soil contributes with greater relevance to the sodium contamination of the bricks used in the buildings located near the research site. The groundwater, despite the greater dispersion in the results, is not ruled out as contributing to the sodium contamination of the bricks. It is also verified that the undamaged bricks arrived from their suppliers without high

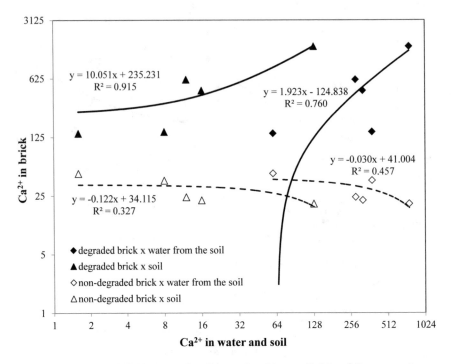

Fig. 4.3 Comparison of calcium results found in samples of damaged bricks relative to groundwater, soil and undamaged bricks

sodium content in their composition, however, it is necessary to analyses the sodium contents brought from the repositories used by the manufacturers, since the existence of this type of element in undamaged bricks is perceptible.

4.5 Chlorides in Atmosphere

The values shown in Table 4.9, resulting from the samples at the five points adopted for the research, represent the chloride content in the atmosphere during a period of 180 days.

Table 4.9 shows that in Petrolina-PE, the chloride content in the atmosphere during six months (sampling period) in the localities adopted for this work was presented as follows:

- PM-1: In the samples collected, the values obtained vary between 11.01 and 41.29 mg/m^{2}day.
- PM-2: Among the values in Table 4.9, the respective values were higher than at other points, ranging from 28.06 to 42.14 mg/m^{2}day.

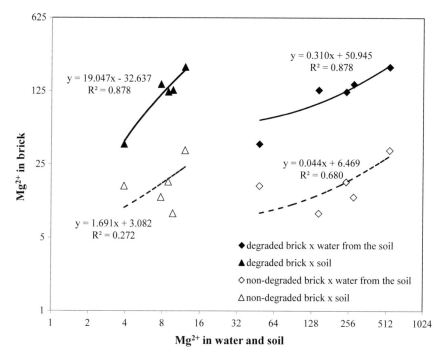

Fig. 4.4 Comparison of magnesium results found in samples of damaged bricks relative to groundwater, soil and undamaged bricks

- PM-3: The values of the analyzed samples presented values ranging from 20.25 to 66.20 mg/m^{2}day, reaching a level considered to be higher in relation to the other points.
- PM-4: the results obtained were very close to those already reported for PM-3, ranging from 21.01 to 53.24 mg/m^{2}day.
- PM-5: Of all the samples analyzed, the lowest value (6.97 mg/m^{2}day) was recorded at PM-5, with the highest registering only 28.33 mg/m^{2}day.

Figure 4.7 shows the results obtained regarding the quantity of chlorides present in the atmosphere at the five points adopted for the research at Petrolina-PE, and the order of magnitude of these results when compared with other results researched and published by Pontes (2006), on the beach of Boa-Viagem in Recife-PE. With these results, it is possible to conclude:

Considering all the values obtained during the six-month survey at the city of Petrolina-PE, PM-3 reports the most representative value of chloride in the atmosphere, registering 66.20 mg/m^{2}day. The lowest value found in this period is recorded in PM-5 (6.97 mg/m^{2}day).

At PM-5, in all samples collected, the levels obtained were lower than the other points surveyed, with the most prominent values appearing in PM-3. As can be noticed, this fact occurs as a result of the site having presented lower values during the

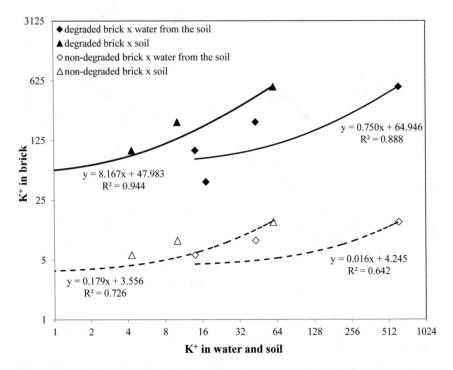

Fig. 4.5 Comparison of potassium results found in samples of damaged bricks relative to groundwater, soil and undamaged bricks

research period in regards to the existence of soluble salts in the soil and groundwater, thus proving that the chloride in the atmosphere varies its value in keeping with the amount of soluble salts in the underground water.

At Petrolina-PE, the chloride in the atmosphere presents values that represent five to twelve times smaller when compared to the data obtained from Boa Viagem. Given that the chloride does not reach the surrounding buildings causing significant damages, its influence can thus be discarded in the process of masonry brick decomposition.

4.6   Critical Analyse

This research verifies that at the lower points of the urban area of the city of Petrolina-PE, the existence of soluble mineral salts is predominant mainly in the soil and groundwater, and that there are low levels of chlorides registered in the atmosphere in relation to values reported by Pontes (2006), obtained at Boa Viagem beach in Recife-PE.

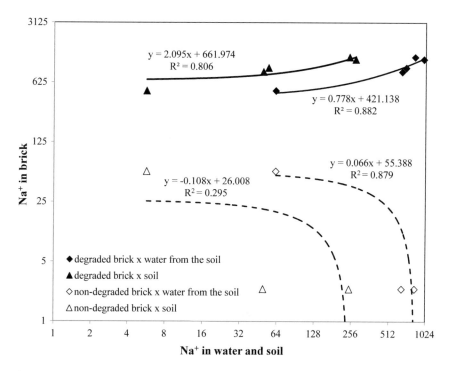

Fig. 4.6 Comparison of sodium results found in samples of damaged bricks relative to groundwater, soil and undamaged bricks

Table 4.9 - Results analysis of chlorides in the atmosphere

Local	Chlorides, Cl^- (mg/m^{2}day)					
	Samp-1	Samp-2	Samp-3	Samp-4	Samp-5	Samp-6
PM1-Raso da Catarina	11.01	12.88	27.18	33.31	29.59	41.29
PM 2-Antônio Cassimiro	31.84	28.06	49.14	42.25	34.70	41.74
PM 3-Jardim Amazonas	20.25	29.36	48.45	37.55	35.43	66.20
PM 4-Dom Malan	21.01	30.49	27.00	37.65	36.77	53.24
PM 5-Vila Eduardo	6.97	10.65	9.65	21.31	26.73	28.33

Based on the auger drillings and monitoring wells of the five research areas, the level of groundwater in the low-lying areas of the Petrolina-PE urban area, which has already been presented, reports a surface depth of 1 m. As shown in the research, it can also be stated that in close proximity to some surveyed areas, although these too are low areas, there is no superficial water table, and that the habitual moisture level is maintained from other contributing sources.

During the implementation of the PM1 to PM5 monitoring wells, the prevalence of frequent humidity with high levels of evaporation was verified in the lower-elevation

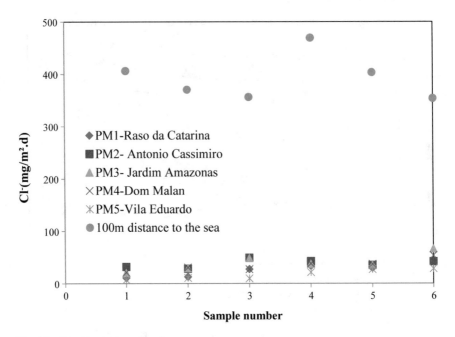

Fig. 4.7 Chlorides in the atmosphere

areas in relation to the contour of the Petrolina-PE urban area. The terrain is characterized by sandy soil at a level very close to the earth's surface, followed by a layer of altered rock, gravel and other impenetrable materials for drilling, which prevents the drainage of water for deeper penetration, thus making the surface saturated, which in turn favors the influx of the moisture into the masonry through the capillarity process.

Soluble mineral salts in groundwater and soil can directly reach building masonry, as well as other parts. The amount of salt in building bricks is not directly proportional to soil, water and atmosphere levels.

Soluble mineral salts present in the soil and groundwater, even though reporting low levels in some places, can achieve access to the buildings through a cumulative salt process over time as a result of the existence of permanent humidity coupled with high evaporation.

The soil is characterized as saline in all the points adopted for the research because they present EC superior to 7 dS/m (at 25 °C).

The calcium and magnesium cations found in the brick samples shows a tendency to receive direct influence from the contribution of raw sewage from their origin in all the surveyed areas, not ruling out the influence of slurry at the locality PM-1.

The raw sewage and leachate, when flooding the soil, do not contribute to increased quantities of exchangeable calcium and magnesium in the soil.

Soil mixed with sewage sludge increases the levels of sodium and potassium in the soil, as well as its electrical conductivity, being one of the factors that contribute to the high levels of these elements found in the researched areas.

Potassium in the soil, its highest level being found in PM-1, may be due to the existence of slurry in the area, in addition to also being influenced by the open-air raw sewage in the area.

The exchangeable sodium present in the soil in PM-3, the highest level in relation to the other points surveyed, is due to the great contribution of raw sewage received from other districts located upstream of that locality.

The electrical conductivity in the soil appears high at all sites surveyed, a fact that can be influenced by the existence of open-air sewage in all locations and slurry at PM-1.

The high levels of the electrical conductivity in the soil characterize the saline soil in all the studied localities.

The percentage of sodium concentrate appears high at all points surveyed, with the main source of influence being the presence of open-air raw sewage in the streets, except for PM-1, which also receives slurry. The high numbers registered of this element characterize the soil as being sodic in nature at points PM1 to PM4 and solodic at PM5.

The sulfate present in the soil at locality PM-1, characterizes the type of soil as moderately serious due to the existing value above 150 mg/l at this locality.

The contamination of the groundwater in the research areas receives contributions from open-air sewers in the streets, and could receive contributions from other sources such as pesticides and existing stabilization ponds in the area. The PM-1, in addition to raw sewage, also receives influence from the existing manure in the sector.

The high electrical conductivity in the water appears very significant when compared to sea water, thus proving that groundwater is not suitable for use in the buildings and its permanence in these researched places directly affects the masonry as well as other parts of buildings.

Some of the sulfate, calcium and magnesium in the groundwater exceed the reference values for sea water, thus proving that the water classifies as saline water with a pH close to eight.

The chloride ion found in the groundwater reports high levels, favoring the appearance of efflorescence in the masonry of the buildings.

The groundwater at PM-e presents higher salinity levels, and is the most critical area in regards to of mineral salts soluble in masonry.

With pH greater than seven at points PM1 to PM4, the groundwater is classified slightly alkaline and a pH less than seven at point PM5 as acidic.

The groundwater collected contained high values of magnesium, calcium, sodium chloride and potassium when compared to sea water.

The groundwater presented a strong degree of aggressiveness in concrete, in parts of the studied localities, when compared to Table 4.7 (see Bauer 2007).

Chloride, sulfate, magnesium, potassium and sodium found in the samples of damaged bricks in most cases exceeded the values found in undamaged bricks, proving that these elements are responsible for the deterioration of the masonry of buildings in the areas surveyed.

The main source of contamination of the damaged bricks points to the groundwater and the soil, as they contained several elements found in the samples of damaged bricks.

The unaffected bricks removed for use as samples at the five points surveyed did not present significant values in relation to the damaged brick samples, showing that the contamination occurred in the location where they were applied in the lower part of the masonry sometime after application, thus exempting the use of brick contaminated by soluble salts to the point of damaging the buildings.

Even in the unaffected bricks not showing signs of aggression from soluble salts, the presence of soluble elements existed in significant levels of chloride, calcium, sulfate and magnesium in relation to the results presented in the damaged bricks, the potassium and sodium intensity being lower.

Damage to the sample bricks occurred through the crystallization of the soluble salts deposited in the masonry, causing the disintegration of particles in the bricks.

The amount of chlorides present in the atmosphere at the Petrolina-PE research sites did not represent significant numbers for the effect of soluble mineral salts in masonry, representing only a variation of 1.9–14% when compared with data obtained in the Boa-Viagem beach in Recife-PE.

The most significant values were recorded at PM-3, where higher values were also present in the groundwater and the soil. The less representative values were observed in PM-5, where the groundwater and the soil also presented lower values.

Chapter 5
Conclusions

The research proved that the presence of soluble mineral salts has great potential to degrade ceramic block in masonry building. The main objective of this work was to analyse the influence of soluble mineral salts on the ceramic brick masonry of buildings, as well as their origins, in the Petrolina municipality area, a region with clay sands and with a great number of open-air water runoff. Clay sands are a type of soil more prone to underground water runoff and, because of this feature, provide a more favourable environment for rising moisture that transports soluble mineral salts to the masonry walls accelerating their degradation process.

The degradation process in ceramic block masonry walls was more influenced by the high content of soluble salts in soil and groundwater than by the amount of chlorides content the atmosphere, but the amount of salt in building bricks was not directly proportional to soil, water and atmosphere soluble salt levels.

The existence of soluble salts in groundwater and surface soil contribute directly to the deterioration process of the masonry, by raising the level of salts by capillarity. However, in hot regions due to more rapid water evaporation and crystallization of soluble salts, these phenomenon's are more worrying as it causes the breakdown of ceramic brick particles and other materials such as concrete, cement mortar and sand.

The properties of durability of fired clay brick walls are very important for the stability and security of masonry buildings constructions over their service life. The research showed that soluble salt transported by natural moisture has potential to generate early deterioration of masonry walls that will affect its performance under normal conditions of use. In this degradation process, the common soluble salts are sodium sulphate and sodium chloride that crystallize within the porous structure of the material creating a pressure that is able to cause the rupture of micro-structure of the material. Special cares should be adopted when salts are present in fire clay bricks or even in mortar used in masonry buildings, mainly sulphates, because they can react with water during the laying of the bricks and a crystallization process may arise with enough pressure to cause damage in wall made with such bricks.

The high electrical conductivity of groundwater plays an important role on this process and measures to minimize such problems, like those presented in the paper,

© The Author(s), under exclusive license to Springer Nature Switzerland AG 2020
J. M. Delgado et al., *Salt Damage in Ceramic Brick Masonry*,
SpringerBriefs in Applied Sciences and Technology,
https://doi.org/10.1007/978-3-030-47114-9_5

should be adopted when it is intended to build masonry structures of ceramic blocks in saline soils.

Chloride, sulfate, magnesium, potassium and sodium ions found in the samples of damaged bricks provide the most important source of degradation for ceramic brick masonry walls.

References

APHA-American Public Health Association, *Standard Methods for Examination of Water and Wastewater*, 22nd edn. (Publication Office, New York, USA, 2012)

L.A.F. Bauer, *Construction Materials*, Volumes 1 and 2, LTC- Livros técnicos e Científicos (2007)

V. Brito, T. Diaz Gonçalves, Drying Kinetics of Porous Stones in the Presence of NaCl and NaNO3: experimental assessment of the factors affecting liquid and vapour transport. Transp. Porous Media **100**, 193–210 (2013)

A.E. Charola, Salts in the deterioration of porous materials: an overview. J. Am. Inst. Conserv. **39**, 327–343 (2000)

E. Doehne, C. Price, *Stone conservation: an overview of current research*, 2nd edn. (The Getty Conservation Institute, Los Angeles, 2010)

E. Doehne, Salt weathering: a selective review. Geol. Soc. London Spec. Public. **205**, 51–64 (2002)

EMBRAPA, *Manual of Soil Analysis Methods*, 2nd Rio de janeiro, Brazil, 212 p. (1999)

M.C.A. Feitosa, Sewage sludge: some applications in engineering. Recife PE, MSC. Thesis, Universidade Católica de Pernambuco, Brazil (2009)

E. Franzoni, Rising damp removal from historical masonries: A still open challenge. Constr. Build. Mater. **54**, 123–136 (2014)

A.S. Guimarães, J.M.P.Q. Delgado, V.P. Freitas de, Rising damp in building walls: the wall base ventilation system, Heat Mass Transf. **48**, 2079–2085 (2012)

A.S. Guimarães, J.M.P.Q. Delgado, V.P. Freitas de, Rising damp in walls: evaluation of the level achieved by the damp front, J. Build. Phys. **37**(1), 6–27 (2013)

S. Gupta, H. Huinink, M. Prat, L. Pel, K. Kopinga, Paradoxical drying of a fired-clay brick due to salt crystallization. Chem. Eng. Sci. **109**, 204–211 (2014)

C. Hall, W. Hoff, Liquid movements. Mater World **15**, 24–26 (2007)

J. Lindqvist, Rilem TC 203-RHM: repair mortars for historic masonry. Testing of hardened mortars, a process of questioning and interpreting. Mater. Struct. **42**, 853–865 (2009)

B. Lubelli, R. van Hees, H. Huinink, C. Groot, Irreversible dilation of NaCl contaminated lime–cement mortar due to crystallization cycles. Cement Concrete Res. **36**, 678–687 (2006)

P.K. Metha, P.J.M. Monteiro, Concrete: microstructure, properties and materials, IBRACON, Editor: Nicole Pagan Hasparyk, São Paulo, Brazil, ISBN: 978-85-98576-21-3 (1994)

G. Moriconi, M.G. Castellano, M. Collepardi, Degradation and restoration of mansory walls of historic buildings—a case history: Vanvitelli's Mole in Ancona. Mater. Struct. **27**(7), 408–414 (1994)

E.Q. Motta, S.R. Ferreira, Variability of compressibility and collapse potential of a source of flood liquid. Revista de Estudos Ambientais **13**(1), 28–41 (2011)

© The Author(s), under exclusive license to Springer Nature Switzerland AG 2020
J. M. Delgado et al., *Salt Damage in Ceramic Brick Masonry*,
SpringerBriefs in Applied Sciences and Technology,
https://doi.org/10.1007/978-3-030-47114-9

S.C. Nappi, L.M. Marques, *Salinity in old buildings, CIMPAR-2010, June 2–4* (Córdoba, Argentina, 2010)

NBR 6211: tmospheric corrosion - Determination of the chloride deposition rate in atmosphere by wet candle method, Rio de Janeiro: Brazil (2001)

NBR 6459: Soil—Liquid limit determination, Rio de Janeiro, Brazil (2016)

NBR 7180: Soil—Plasticity limit determination, Rio de Janeiro, Brazil (2016)

NBR 7181: Soil—Grain size analysis, Rio de Janeiro, Brazil (2016)

NBR 13553: Soil-cement materials for monilithic walls of soil-cement without structural function: Requirements, Rio de Janeiro, Brazil (2012)

NBR 13602: Soil—Dispersive characteristics of clay soil by double hydrometer - Method of test, Rio de Janeiro, Brazil (1996)

NBR 15495-1: Monitoring wells of ground water and granular aquifers Part 1: Design and construction, Rio de Janeiro, Brazil (2007)

NBR 15847: Ground water sampling in fwells - Purging methods, Rio de Janeiro, Brazil (2010)

L. Ottosen, I. Christensen, Electrokinetic desalination of sandstones for NaCl removal—test of different clay poultices at the electrodes. Electrochimica Acta **86**, 192–202 (2012)

J.M. Paz-García, B. Johannesson, L. Ottosen, A. Ribeiro, J. Rodríguez-Maroto, Simulationbased analysis of the differences in the removal rate of chlorides, nitrates and sulfates by electrokinetic desalination treatments. Electrochimica Acta **89**, 436–444 (2013)

L. Pel, H. Huinink, K. Kopinga, Ion transport and crystallization in inorganic building materials as studied by nuclear magnetic resonance. Appl. Phys. Lett. **81**, 2893–2895 (2002)

Petković, H. Huinink, L. Pel, K. Kopinga, van R. Hees, Salt transport in plaster/substrate layers, Mater. Struct. **40**, 475–490 (2007)

R.B. Pontes, Dissemination of chloride ions on the sea front of Boa Viagem, Recife—PE, MSC. Thesis, Universidade Católica de Pernambuco, Brazil (2006)

H. Prado, Soils of Brazil: Genesis, morphology, classification, survey and management. 4th Edition, 281p., Piracicaba-SP, Brazil (2005)

A. Putnis, M. Prieto, L. Fernandezdiaz, Fluid supersaturation and crystallization in porous media. Geol. Mag. **132**, 1–13 (1995)

I. Rörig-Dalgaard, Further developments of a poultice for electrochemical desalination of porous building materials: minimization of side effects. Mater. Struct. **48**, 1901–1917 (2015)

I. Rörig-Dalgaard, Development of a poultice for electrochemical desalination of porous building materials: desalination effect and pH changes. Mater. Struct. **46**, 959–970 (2013)

A. Sawdy, A. Heritage, L. Pel, A review of salt transport in porous media: Assessment methods and salt reduction treatments: Assessment methods and salt reduction treatments, Salt Weathering on Buildings and Stone Sculptures, 22–24 October 2008, The National Museum Copenhagen, Denmark, 1–27 (2008)

G. Scherer, Stress from crystallization of salt. Cement Concrete Res. **34**, 1613–1624 (2004)

G. Scherer, Crystallization in pores. Cement Concrete Res. **29**, 1347–1358 (1999)

I.TS. Silva, Identification of the factors that cause efflorescence in Angicos buildings, MSc. Thesis, Universidade Federal Rural do Semiárido, Brazil (2011)

L.C.A. Tavares, The Question of social housing: challenges and perspectives. Jus Naviganti, Terezina-PI (2004). Acessed 10/06/2018, http://www1.jus.com.br/doutrina/texto.asp?id=5396

V. Voronina, L. Pel, K. Kopinga, Effect of osmotic pressure on salt extraction by a poultice. Constr. Build. Mater. **53**, 432–438 (2014)

G. Zappia, C. Sabbioni, C. Riontino, G. Gobbi, O. Favoni, Exposure tests of building materials in urban atmosphere. Sci. Total Environ. **224**, 235–244 (1998)

D. Young, D. Ellsmore, Salt attack and rising damp: a guide to salt damp in historic and older buildings, [2nd ed.], Heritage Council of NSW, Heritage Victoria, South Australian Department for Environment and Heritage, Adelaide City Council, Australia (2008)

T. Warscheid, J. Braams, Biodeterioration of stone: a review. Int. Biodeterioration **46**, 343–368 (2000)

Printed in the United States
By Bookmasters